# MY 1st BOOK OF AI

# Rainbow Arts Studio

Copyright © 2026 Sara Kale
All rights reserved.

All copyright in this publication and its contents is reserved by the author, and no part may be reproduced, stored in a retrieval system, or transmitted in any form or by any means without the author's prior written permission. However, critical reviews and certain non-commercial uses allowed under copyright law are permitted.

Website: https://rainbowartsstudio.com/
Amazon Page: amazon.com/author/sarakale
Email: sara.rainbowartsstudio@gmail.com
IG: rainbow_artsstudio

# What Is AI?

Have you ever asked a phone a question?

Or watched a video chosen just for you?

That was AI!

**AI stands for Artificial Intelligence.**

"Artificial" means made by people.

"Intelligence" means the ability to learn from patterns.

AI is a computer program.

It learns from millions of examples.

Then it uses what it learned to help us.

AI is a tool. People build it.

People use it. And people are always in charge.

# AI Is a Tool, Not a Person

A hammer helps you build.

A pencil helps you write.

AI is a tool too, but it lives inside computers
phones and other devices.

AI does not have feelings.

It does not get hungry, tired, or sad.

It does not love you the way your family does.

AI only does what people tell it to do.

People build it. People fix it. People turn it off.

Knowing this helps you use AI the smart way!

ON
OFF

# How You Learn

Think about how YOU learned to read.

First, someone showed you letters.

You practised sounds.

You made mistakes. You tried again.

Your brain got better slowly and beautifully.

You also learn from touching, tasting, seeing, and hearing.

You learn from your family, your friends, and from playing.

Your feelings help you learn too.

When something is exciting, your brain pays extra attention.

You are an amazing learner.

# How AI Learns

AI learns very differently from you.

It does not play at recess.

It does not hug its family.

AI learns by looking at hundreds and thousands of examples.

To teach AI to recognise cats, you show it thousands of cat photos.

It finds patterns. And next time, it can spot a cat!

But what if AI sees a picture it has never seen before? It makes its best guess. Sometimes that guess is wrong.

A dog wearing a cat costume might trick it!

That is why people always check AI's work.

CAT
SING
?

# SORT LIKE AI with Leo and Maya

Leo and Maya tried this. Now it is your turn.

You will need: pictures of animals, vehicles, and foods. Mix them all up.

- Spread them on the floor, just like Leo did.

- Sort them into groups. You decide how.

- Look at your groups. How did you decide what belongs together?

That is how AI learns. It looks at pictures and figures out what goes together just like you did.

Now try again. Sort by color, by size, and by how they make you feel.

The more ways you try, the more you think like AI!

# Your Amazing Senses

Close your eyes. What do you hear?

Take a breath. What do you smell?

Touch your sleeve. What does it feel like?

You have five amazing senses.

Sight. Hearing. Smell. Taste. Touch.

Your senses send messages to your brain all day long.

And your senses connect to your feelings!

The smell of cookies makes you feel warm.

The sound of thunder might make you feel nervous.

Your senses make you, YOU.

# AI's Senses

AI does not have a nose or ears like yours.

But it has tools that help it take in information.

A camera helps AI "see."

A microphone helps AI "hear."

Sensors can measure temperature and movement.

But AI does not FEEL anything.

It does not enjoy the smell of rain.

It cannot feel excited or surprised.

AI collects information.

Your senses create experiences.

That is a very big difference!

1235
3 . 49

# A Short History of AI

People have dreamed about thinking machines
for a long time.

In 1950, a scientist named Alan Turing asked:
"Can a machine think?" That question started
everything!

In the 1980s, computers learned to play chess.

In the 2000s, phones learned to understand our
voices.

Today, AI helps doctors, farmers, scientists, and
kids!

AI is still very young.

And here is the exciting part.

People like YOU will help write what comes next.

# AI Around Us Today

AI is already part of your day. And you might not even know it!

When your map app finds the fastest road, that is AI.

When a video site picks your next show, that is AI.

When a smart speaker answers a question, that is AI.

At hospitals, AI helps doctors read X-rays.

On farms, AI helps grow healthy food.

Some AI can even write stories and draw pictures.

AI is everywhere, helping people work better.

But people always make the most important decisions.

Always.

SCHOOL
recommended for you

# Your Brain vs. AI

Your brain and AI are both good at learning.

But they work in very different ways.

Your brain can smell a garden, feel excited about a birthday, or know when a friend needs a hug.

AI cannot do any of that.

AI can search millions of pages in one second.

Your brain cannot do that.

But here is the most important thing.

Your brain has kindness, curiosity, and creativity.

No computer has ever had those.

And that makes YOU very special.

search

# People Guide AI

AI does not decide on its own what to do.

People give it rules. People test it. People fix it.

Scientists help AI get better at finding patterns.

Engineers build the systems.

Teachers help children learn to use AI wisely.

For big decisions, people always check AI's work.

Smart users ask questions too.

When AI says something unfair, a person fixes it.

You have a role!

If an AI answer seems wrong, tell an adult.

That is how we make AI better. Together.

# When AI Is Helpful

AI can be a wonderful helper when you use it the right way!

It can help you find information quickly.

It can read text aloud if that helps you learn.

It can help doctors find problems early.

It can help scientists study weather and protect animals.

But the best helpers work WITH people, not instead of people.

Your creativity and kindness make every tool work better.

Remember. AI is the helper. YOU are in charge.

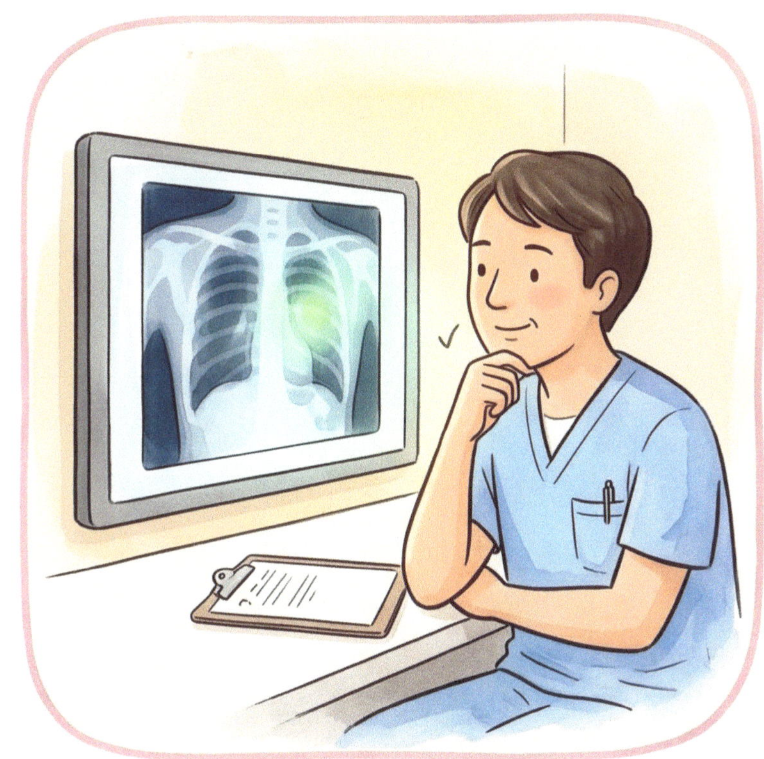

# Being Careful With AI

AI is useful. But it is not perfect.

AI can get things wrong and can sometimes say things that are not true.

Always ask a trusted adult before using a new app, website, or camera.

Your private information is yours to protect.

If something feels wrong or scary, stop and tell an adult.

Smart users ask questions. They check answers.

They talk to people they trust.

You are in charge. That is a big, important job.

# Smart Habits for Kids

## MY AI EXPLORER BADGE

Check every habit you practice!

☐ I ask a trusted adult before using a new app or website.

☐ I check if what AI says is true. I ask an adult or look in a book!

☐ I keep my information private. My name, address, and photos are mine to protect.

☐ I take screen breaks. I play outside. I spend time with people I love.

☐ If something feels wrong, I speak up. I ask for help.

Check all five? YOU are an official AI Explorer!

# Your Family and Friends Matter Most

AI can feel helpful. But it is not your friend.

It cannot love you, worry about you, or give you a hug.

Real friendships are built by spending time together, sharing feelings, and truly caring about each other.

AI cannot do any of that.

Put down the screen. Run outside. Call a friend.

Draw something. Laugh with your family.

Your family and friends are the ones who truly know you.

They are your heart.

AI is just a tool.

Your people are your world.

# Note for adults

Thank you for reading My First Book of AI with your child. This book was written because children today are growing up alongside artificial intelligence in their homes, classrooms, and on the devices they use every day. Rather than waiting for children to encounter AI without context, this book gives them the words, ideas, and questions they need to engage with it thoughtfully.

What this book addresses:

Privacy and safety. Children are encouraged throughout this book to always ask a trusted adult before using any app, and to treat their personal information as something precious and private. This is not a book that promotes unsupervised AI use.

Emotional attachment. Spread 14 (**Your Family and Friends Matter Most**) directly addresses the risk of children treating AI as a friend or companion. The message is clear and loving. AI is a tool. Your people are your heart.

Over-reliance and critical thinking. Multiple spreads teach children that AI makes mistakes, that humans must always check AI's work, and that their own curiosity and creativity are what make any tool truly powerful.

Screen time and real life. This book does not celebrate screen time. It closes by celebrating running outside, calling a friend, drawing, and laughing with family.

What this book is not. It is not AI hype. It does not suggest AI will replace teachers, parents, or human connection.

Best practice. Read it together. Pause at each spread. Use the Conversation Starters below. Let your child ask questions you do not know the answer to, and look them up together.

# Conversation Starters

### For home or classroom

1. What can you do that you think AI could NEVER do?

2. If you did not know the answer to something, would you rather ask AI or ask a person, and why?

3. How would you know if an AI answer was wrong? What would you do next?

4. Can you think of one rule our family or class should have about using AI?

5. What is your favorite thing to do that has NOTHING to do with a screen?

6. If an AI said something that made you feel uncomfortable, who would you tell?

7. What do you think AI will be able to do when you are an adult? What do you hope PEOPLE will still be doing?

# Source Notes

Spread 1 (What Is AI?): Russell & Norvig (2020). Artificial Intelligence: A Modern Approach (4th ed.). Pearson.

Spread 4 (How AI Learns): LeCun, Bengio & Hinton (2015). Deep learning. Nature, 521, 436-444.

Spread 7 (History): Turing, A.M. (1950). Computing machinery and intelligence. Mind, 59(236), 433-460.

Spread 8 (AI Today): U.S. FDA (2023). AI/ML-Based Software as a Medical Device. www.fda.gov.

Spread 10 (People Guide AI): AI4K12 Initiative (2021). Five Big Ideas in AI. Carnegie Mellon. ai4k12.org.

Spreads 12-13 (Safety): Common Sense Media (2023). AI Literacy for Kids. commonsense.org.

www.ingramcontent.com/pod-product-compliance
Lightning Source LLC
Chambersburg PA
CBHW042128030726
47599CB00002B/384